"If you are looking for a well-spoken, man- utmost integrity, Matt Montgomery is the ranchman you can count on. He knows what it takes to manage a ranch and keep it looking presentable, and he knows the true value of covering all the details from start to finish. From my first meeting with Matt, I believed his pledge that not a day would go by that he and his team were not brainstorming about how to market and sell my ranch. Matt Montgomery was at every showing, was an out-of-the-box thinker, and showed the property to its highest and best use at all times. The hustle that Matt and his crew exhibited was nothing short of magic, resulting in dozens of showings, a tremendous social media presence, and multiple offers. More than selling a lovely ranch to a delighted new investor, Matt will remain a referral partner and has earned my respect and friendship for life."

—Kellie Rodenbeck,
Branch Manager, LeaderOne Financial Corporation

"Matt's knowledge of land and property is only surpassed by his passion for sharing it. He understands how to care for properties so that you and your family can enjoy your property to the fullest."

—Harrison Thomas,
Rocker U Ranch

"It was our delight to have Brazos Ranch Group manage our property in Chappell Hill. Matt has been a competent partner for advice, method, and maintenance as well for machinery updates and repairs. We always feel confident with his team and service."

—Dougal and Cathy Cameron,
Los Arboles Farm

"This past year has been so much fun, and we are so excited for our new peaceful life in the country, entertaining family and friends. We know, without a doubt, that Matt had the biggest hand in making it all possible for us. We are so thankful for you, Brazos Ranch Group!"

—Gerrald and Marianna Giblin,
Brenham, Texas

"We bought our farm three years ago. We immediately were inundated with things to do, papers to fill out, and contractors to find. We constantly called our friends who had places, but they seemed to be struggling with the same things. I mean this 100% when I say, 'Then our Knight in Shining Armor arrived.' If we hadn't met Matt and his team, we would have sold our ranch by now. He knows the right people for the job, and I highly trust him with my place."

—Bethany C.,
Round Top, Texas

THE WEEKEND RANCHER

AN EXPERT'S GUIDE TO BUYING, MANAGING, AND LOVING YOUR COUNTRY PROPERTY

MATT MONTGOMERY

LUCIDBOOKS

The Weekend Rancher
An Expert's Guide to Buying, Managing, and Loving Your Country Property

Published by Lucid Books in Houston, TX
www.LucidBooksPublishing.com

ISBN-10: 1-63296-324-8
ISBN-13: 978-1-63296-324-6
eISBN-10: 1-63296-293-4
eISBN-13: 978-1-63296-293-5

Special Sales: Most Lucid Books titles are available in special quantity discounts. Custom imprinting or excerpting can also be done to fit special needs. Contact Lucid Books at Info@LucidBooksPublishing.com.

Table of Contents

Introduction 1

Chapter 1: Choosing the Perfect Place 3
Chapter 2: The Problem 9
Chapter 3: Finding Good Help 13
Chapter 4: Safety 19
Chapter 5: Equipment 23
Chapter 6: Must-Haves for Your Ranch 27
Chapter 7: Investment 33
Chapter 8: Wildlife Tax Exemption 39
Chapter 9: Regrets 43
Chapter 10: Plugging In 47
Chapter 11: Let's Do This 51

Appendix A: Inclusive Contractor Needs Form 53
Appendix B: Small Equipment List 55

Introduction

Imagine leaving rush-hour traffic behind on Friday afternoon and a short time later pulling into your sprawling ranch miles away from city noise and distractions. Since you've picked up this book, you are likely either ready to purchase your first property, or you've already purchased one and are looking for advice on how to manage it.

If you're like me, you love country living and the escape it provides from the frantic pace of Monday through Friday. You've worked hard and are finally able to put some of that hard-earned money toward creating a weekend getaway where you can spend more intentional time with family and friends. And let's face it, brushing up on your frisbee-golf or skeet-shooting skills sounds like a great way to pass the time.

Some of my favorite family memories are times spent at our weekend ranch. But it does have a unique set of challenges—knowing what equipment to buy or avoid buying, finding good help for standard property maintenance, and setting your property up to offset some of your costs. In this book, I offer advice I've learned over years of owning, managing, and selling countless properties. It's advice to help you avoid expensive mistakes, minimize your stress levels, and get you well on your way to some R&R.

I have a passion for selling and maintaining weekend ranches. As a fellow ranch owner, I want everyone to experience the joy that comes from owning a place in the country. If you're in the market for a recreational property, I'd love to meet with you to talk about your dream place. Let me help you turn your dream into a reality.

BUILDING
BANK
CHAPPELL HILL
BANK

Chapter 1
Choosing the Perfect Place

There are many good reasons to buy property. Before you start looking for the perfect place, it's smart to think through a few specifics. The first and most important is to list the main goal for your property. Is it a tax incentive or write-off? Is it a family recreational property? Is it for your hunting, fishing, or birding hobbies? Maybe you want to own an Airbnb ranch. Do you need a place to work with no distractions? Is it strictly an investment decision? Maybe you have just sold a large investment and are looking to put your profit into another investment through a 1031 exchange. Maybe you want to do all of the above. Defining your top three or four desired goals will be helpful when beginning your search.

Investment

Spending time unplugged on my weekend property was the best financial decision I made this year. I own plenty of property, but I lease a place in West Texas on the Devils River. I checked my calendar, and over the past 12 months, I have spent 23 nights there. Wow! That's nearly a month out of the year. The best investment a person can make is in personal growth. I relaxed, read more than a dozen books, and hiked hundreds of miles. On those hikes, I thought about my business, my investments, my faith, my family, and how to improve myself overall.

I will never forget the longest hike I took on the ranch. My watch says I walked 38,000 steps, but I don't remember any of it. I was deep in my head, thinking about what I could do to be better or different than my competition. Realtors are like mosquitoes—there are tons of them swarming around. On this particular hike, I developed a list of practices I wanted to implement to maintain a competitive advantage. I have seen the fruit of the ideas generated on that hike bring me monetary gains that will pay for the lease at Devils River for years to come. Investing in real estate can be a profitable venture, and a large part of that may not be from selling it. I'm a believer in owning a place you can work on and also unplug.

Buying property is an investment on a number of levels. If you purchase a property correctly and maintain it well, it is a solid, long-term investment simply because of appreciation. Historically, land prices rise and fall, as in any market. But because of land's guaranteed scarcity, it is very likely to continue to rise in value over time. We've all heard our grandmas say, "You can't make more land!"

You'll be able to write off many of the activities that take place on your property, and thanks to agricultural exemptions, you'll likely pay very little in taxes each year, unlike investing in homes or commercial property. **If you are looking for a place to simply park your money, I believe raw land is a strong option to consider.** Keep in mind that every structure you add is taxable and not under the umbrella of an agricultural exemption. I always recommend that my clients buy more developable acres and fewer infrastructures that result in long-term expenses, much like the house you live in now. In a world of unstable markets, land has a lot of potential upside. I always buy as much land as I can and then lease it for cattle, hay, or hunting. I also lease the structures on the property. My goal in buying a piece of land is to offset the payment and bide my time for a very likely appreciation. Wise people own large tracts of land.

Lending

There are several great local banks near me that understand current values and thrive on giving loans for land. A small-town local bank near the land you are purchasing is a smart way to get a loan. These banks know the community, the track record of appreciation, and likely some of your neighbors. Local bankers also become great resources for the area norms; qualified contractors; and a list of great places to relax, eat, and drink. They also tend to know whom to avoid. If you are in Texas, reach out to my banker at Citizens State Bank in Brenham at cflencher@csbtx.com.

The other lending option for raw land is through farm credit lenders. They are agricultural lenders whose rates are ultra-competitive. You become a part owner, and your effective interest rate is lowered to a fair market value based on the profits of the co-op. These loans normally require 20 percent down, whereas a local bank may be able to get you in a property with a much lower down payment if it appraises high enough.

Many of the raw land deals that have fallen through on my watch were the result of large corporate banks that don't know or understand local values or have so many hoops to jump through that it's nearly impossible to close a deal. The value in using local in this instance pays for itself in relationships made and local information given. **I highly recommend using a bank near the property you are purchasing.**

Hobbies

Before you purchase your property, it is important to decide what activities you most want to do when you're there. That is where you can let your inner child come out. Our company specializes in recreational property, which means we know how to filter properties so you are matched to the one that best fits your wants. It is impossible to find a place that is 100 percent what you want, but certain attributes are key to meeting your needs. Be sure to study the map of properties surrounding yours to ensure that nothing will hinder the activities you are most excited about. For instance, peace and tranquility is a must, so buying a place with a weekend racetrack next door would be a nightmare. I have seen a beautiful country property literally right next door to a drag strip where cars raced every Saturday night. I have also known people whose biggest desire was hunting or fishing look at places without any trees or water.

Here are some of the hobbies and attributes many of my clients are looking for on their ideal property: hunting, fishing, scenic views, horses, silence, skeet shooting, and hiking. Some want to be near their kid's college, and most important, they want the property to be family-friendly. All of these are possible with the right place, but be cautious how you prioritize them and be willing to be flexible on a few.

Distance from Home

How far is the property from your home? This is worth examining for many reasons, including investments. Obviously, the closer the property is to a large city, the more expensive the acreage will be. It is shocking how much more land you can get by driving an extra 30 minutes. I have many clients who say they want to be one hour from Houston, but once we look around and they discover they can have twice the land if they go 25 or 30 minutes farther, they change their minds.

Here is a tip I've learned over the years. What's most important is not how far the ranch is from your house but how close it is to a town that has reasonable amenities. Driving two hours after work on Friday to your ranch isn't inconvenient if once you arrive, there are restaurants and grocery stores within a 20-minute radius. I have heard from clients that driving from their house to the ranch provides them the perfect opportunity to unwind and leave work behind.

Budget

I will never recommend overextending your budget for a farm or ranch, but I will share how owning a place has personally made me more money than it will ever cost me. The ability to turn off my phone, sit at the picnic table under my pecan tree, and work on my career has paid for itself time and time again. Be prepared to talk budget with your real estate agent immediately and be realistic with the numbers. There is no pride or embarrassment in having a smaller budget than the next guy.

Spend time thinking about the goals you have for your property, and when it's time to start looking, have your top two or three prioritized. Remember, be flexible, but I'm confident that the ideal place is waiting for you.

The Perfect Place Testimonial

Jeff sent me a text message recently that he was sitting on his dock, reading a book, catching the occasional bass jumping out of the water in his periphery when he realized he had the perfect place. Our team manages his place with our unique method of "perfect upon arrival," and he utilizes it a couple days a week to work on his business. He has told me numerous times that the cost of his ranch will pay for itself 20 times over in his productivity and effectiveness in his business. The time to unwind, journal, and think without distraction is priceless. Yes, it costs money to own a ranch, and yes, a person can buy and self-manage, but what is the cost of the time spent? Could you double your income from a peaceful place to work? Could your family grow closer without the distractions of phones or TV? Could you engage in conversations with your kids over a day of fishing—conversations about their innermost struggles and fears or their greatest joys?

Chapter 2
The Problem

One of the challenges I hear most often from weekend ranchers is that owning and maintaining two separate households is more difficult than they expected. **Setting up systems is the key to peaceful ownership and something we can help you with.**

Why It's Hard

While this won't be the case for every family, I'm going to lay out a typical scenario that comes up for many weekend ranch and farm owners, particularly families with young children. The husband works hard all week at his day job, which affords them the ability to buy the weekend ranch. After an intense workweek, he would like to enjoy some downtime. When Friday rolls around, everyone is excited about getting away to the ranch.

The problems start when the family finally arrives at their home away from home. Instead of kicking back, you watch as your wife starts sweeping bugs, wiping down counters, and running loads of sheets through the laundry because they feel damp and musty. While you may not mind or even notice these issues, your wife wants to stay in a house that meets her level of cleanliness. If you have infants or toddlers, the floors need to be swept and mopped before she can feel good about letting them roam around. Your wife starts to feel like this weekend away for R&R is actually a lot more work than if the family had just stayed home. And before anyone knows it, it's Sunday afternoon and time to strip the beds, pack up the food, take out all the trash, and make sure everything is cleaned up so the place looks decent when you come back. But let me assure you, there is a better way.

An obvious solution is to hire a housecleaner who can clean the day before you arrive and then again after you leave. But that creates its own set of tasks such as locating a good cleaner, setting the schedule, making payments, and managing any issues that come up pertaining to house upkeep. You don't want to walk into a shut-up house a week after the fridge stopped working. The half-honest guy you hired to come take a look ends up charging more than he quoted. The stress level is skyrocketing, and the amount of time spent managing the ranch is becoming more than you ever imagined. The dream is becoming a nightmare.

We can help you with this. **Let us handle hiring and managing good people who will show up like clockwork and do their jobs well.** Be willing to pay $20 per hour to hire a worker who can make sure the house is ready when you arrive and will clean it after you leave. It may cost a little extra, but think of all the hours it will free up.

The goal is to have a family space where everyone gets a chance to relax on weekends. After all the hustle and bustle of Monday through Friday, you and your family deserve a chance to hang out by the firepit, play a round of frisbee golf, and enjoy cooking a delicious meal together. That is why you bought the weekend ranch to begin with. If you find ownership too onerous or stressful, you may be trying to do too much yourself. Let us help you manage hiring good help so you have time to do what matters most.

The Problem Story

I often get hired to clean up properties so they can be sold. Owners buy places, and the husband gets all the cool new equipment, a sweet new set of boots, and hits the pastures running. His wife starts planting vegetation and trees to match their lavish gardens at home. They find it so fun to get out of work on Friday, head to the ranch, and start planting, mowing, and DIYing the ranch house. Then they get animals—five horses, 37 chickens, a mutt, and a donkey named Jack. It's a blast. Then, they build a barn, a skeet range, and a new main house. They were able to mange the place alone at first, but they have added so much stuff that they need help—and fast. They have created two or three full-time jobs on the ranch. But for the dream to stay a dream, they need to manage their expectations and *get help*. I am happy to use my team to clean up and sell a property, but I also love it when we do the work and send them a bill and they decide to keep their dream place and give it another try.

Chapter 3
Finding Good Help

Helping clients find good help is probably how I can help the most. We have a great pipeline of hardworking individuals who work long hours on ranches and enjoy it. But country life moves at a slower pace, and many new-to-country-life clients become frustrated when someone they've hired doesn't answer the phone or text back quickly. City life is full of technology, making communication between people practically instant. So, it is sometimes a struggle for new property owners to adjust to the slower pace of the country. But my workers labor at a steady pace, are trustworthy and professional, and will knock out your ranch to-do list.

Methods for Managing Ranch Work

There are several ways to handle the work on your new property. One method is to buy very little of your own equipment and then hire individual contractors—housecleaner, yard company, arborist, field maintenance shredder—to do specific tasks. This approach costs more on a monthly basis and is less customizable to the landowner, but it is often a good place to start. The benefit of this method is that it allows you time to figure out what equipment you need; understand what work needs to be done; and find good, long-term workers. The downside is the cost, the inability to schedule workers around your weekend trip, and the lack of personal contact with contractors.

Depending on the size of your property, it is common to find a neighbor who has a full-time employee he or she would allow to come over in the evenings and on Saturdays to knock out your tasks. This can be a great approach because the helper is already working on a ranch and understands the processes. It's also an efficient way to share expenses with someone else rather than having someone full time on just your place. **A common misconception among property owners is the amount of time it takes to keep up the grounds.** I have found that a decent-sized weekend ranch can easily sustain one or two people full time with occasional help from a housecleaner. Sharing workers with the neighbor is a smart way to start.

If your property has the infrastructure, another reasonable approach is to have an individual or couple live in a small cottage on your property, receive a modest salary, and, in exchange, upkeep your property. This method has great potential to create a win-win for both parties. It's reasonable to think the individual or couple could hold a full-time job and do the upkeep tasks on evenings and weekends. It is hard to get help with just the cottage as compensation, so normally, some monetary payment is needed. When I show properties, I look at the structures and talk with my buyers about the possibility of using one of the smaller structures as a home for a maintenance couple. Having someone keep an eye on your property and be available when you have issues is invaluable.

If you're buying raw land and building your dream home with a pool, recreational games, and landscaping, you should be aware of the future costs. Yes, it is worth it, but it's also good to know what you're getting into. We maintain a very nice, always-ready weekend property. Our budget would blow most people's minds, but the reality is that if you have a huge house that is not your primary residence and plenty of farm infrastructure or improvements, it is costly to keep it ready for your arrival. If you have dreamed of adding all of these features and you do not budget for maintenance expenditures, these added features can quickly look trashy and unusable.

Solutions

Connect with me on my website or give me a call. I would be happy to talk you through expected expenses based on what you envision for your ranch. I have 20 to 30 full-time ranch workers, some of whom I am able to share with nearby ranchers and definitely with my real estate clients.

I have received loads of positive feedback about the productivity and high-quality work my workers do. A lot of people are stumped by this, but finding quality people to do ranch work has been easy for me. I am definitely a detail person, and I expect the quality of work they do to meet my high standards, but I also love each and every one of them. It is always my goal to help them improve the quality of their lives and their families' lives. One of the biggest challenges for a weekender is forming a relationship with their workers that creates trust and results in their helpers wanting to achieve the best possible outcome for the owner.

The owners of the ranches we manage take time to get to know me and my family, and we strive to give them 110 percent and always make the weekend trips to their properties spectacular. When interviewing or planning to hire a person to work your place, think strategically about how you can improve their lives and use your connections to help them continue to grow in their professions. Once you have found a ranch helper you love, be sure to make that person comfortable and help him or her grow as an individual.

Contractors

Appendix A presents an inclusive contractor needs form. It is priceless if you are in the process of buying a property. It lists every supplier you could possibly need—electrician, plumber, water well guy, septic guy, and so on. Hopefully, you'll get two options for each service. If these providers have been to the property before, call them and introduce yourself as the new owners. If they aren't familiar with your property, tell them the address and let them know you will be calling them to bid on work you have in the future. It's also helpful to take this list to your neighbors and the people selling you the ranch and get their contact information.

If you find yourself wanting to try a local contractor who wasn't recommended, it's a good idea to interview them first. I create a task list and hire a new company to do small tasks first and then watch how they perform before I give them a large project. I hire them to build the well house or equipment shed, and I look at the quality of their work before I set them free to build my house. That will help you avoid possible disasters with companies that oversell what they are capable of. Get lots of references and call, visit, and study them.

Having trustworthy, qualified contractors for specialty work will give you peace of mind. We provide this list for clients who buy property through us or whose property we manage. If there aren't enough vendors on your list, another option is to contact your local chamber of commerce or the Better Business Bureau, or even ask your favorite electrician from the city to come out to your farm and do the task. If you have a great relationship with someone, it often works best to allow them to come out to your property. The main advantage of using contractors who already know your property is that they are familiar with all your systems and where items are in your yard and home.

Finding skilled workers to help you on your ranch is essential. It saves you time and money, and when using local help, it generates jobs for the town's economy. Take time to build a relationship with your contractor or vendor. Country living is more like pioneer living and demands more reliance on your neighbors and the people in the community. Helping others and letting them help you is part of small-town life. It generates loyalty, higher productivity, and a stronger work ethic, which all contribute to a well-managed ranch, a strong local community, and the life you pictured when you bought your place.

Chapter 4
Safety

Generally speaking, weekend ranch ownership has the potential to be one of the most fun, relaxing ventures you will undertake. But understanding the risks and taking safety precautions are critical. This chaper is the one cautionary part of this book. None of the content is an attempt to scare you but rather to open your eyes to some potential dangers.

I have to admit, I do have several horror stories of accidents that resulted from circumstantial irresponsibility. With that being said, on all the ranches I have worked, we have had no serious injuries. Many things I will caution you against have never happened to me, my team, or 99 percent of the ranch owners I know.

Safety Gear

You will likely own some cool equipment you've always wanted but have no real experience using. It is very important to take all precautions to protect you and your family while using the equipment. **Wives, here is a list of necessary safety items to buy for your husband before he goes out to play with his new chainsaw and tractor.**

- *Gloves* – Buy heavy-duty leather gloves, sized correctly. Make sure the gloves fit like, well, a glove.
- *Safety glasses* – God only gave us two eyes. Protect your eyes. Sure, go ahead and get the fashionable safety glasses, but just be sure to wear them when you're doing anything where dust or metal is flying.

- *Ear protection* – Most heavy ranch equipment makes a ton of noise. Once again, you've only got two ears. Have you ever noticed that most of the time, old farmers can't hear a thing? They've lived their lives providing for their families using farming equipment, and over time, the high-decibel noise has whittled away their hearing. Keep a pair of earplugs in several locations on your ranch, particularly where you do the noisiest work, so you don't have any excuse *not* to wear them.
- *Proper footwear* – One mistake I frequently see new ranch owners make is the shoes they and their family wear around the property. It is not okay for your kids to roam around the back 40 wearing Crocs or sandals. We kill snakes all the time on our properties. We have never had a family member bitten by a snake, so it's obviously not common, but I am regularly cautioning my family to wear close-toed shoes, preferably boots. You should set the example and do the same when you are working or walking in unmanicured areas. I wear tennis shoes all over my properties, but I carry a pair of rubber boots in case I need to walk through wet areas or tall grass. The one lesson I've learned is that if the area looks like it could have snakes in it, you can pretty much guarantee that it has snakes in it.

Preparedness

It's smart to be prepared on the rare occasion that someone does get injured on your property. Here are some tips for being prepared.

- Locate and take a drive to your local hospital. In an emergency, it's not always easy to think clearly. It is helpful to know exactly where the hospital is located as well as how to get there. It never hurts to call the hospital to find out about the facility, such as whether they perform CT scans, have antivenom on hand, and so forth. It's also a good idea to look at the hospital's ratings online. We have three hospital options, all roughly the same distance from our ranch. We now know the specialties of each hospital and can determine which one to go to based on the particular injury.

- Hang an instructional note in your house and barn so guests have quick access to important numbers, addresses, and contacts.
- Keep a first-aid kit nearby. You will have thorns to extract and bee stings to soothe. Your farm is not like your city house. In the event of a 9-1-1 emergency, the response time is significantly longer. I don't say that to scare you but to help you be mentally and physically prepared. For instance, we had a false alarm in our house and had to call 9-1-1. The ambulance took 25 minutes to arrive. It's good for the local police department, fire department, and volunteer fire department to know who you are and where your place is. **I highly recommend supporting your local volunteer firefighter group and attending their yearly parties and public events so they have a personal connection with you, respect you, and are more prepared to help if you ever need them.** They can also be great contacts for your many ranch needs.

Don't Be Stupid

I love equipment. I own equipment. My kids ride with me and are learning to drive our equipment. But we know the risks and don't allow situations to unfold that put me or them at risk. Don't use equipment you don't understand. Ask as many questions as it takes for you to get comfortable with a piece of equipment. Don't treat your equipment like a toy. Understand its limitations and risks and become proficient enough to avoid all risk.

You should own the tractor you've always dreamed of owning and the utility vehicle that makes getting around your property a breeze. There's no therapy that compares to clearing your mind while mowing the pasture on your tractor, planting the garden, feeding the animals, or watching the sun set on the hills behind your place. Use maturity and common sense, and the dream will stay a dream instead of turning into a nightmare.

Protect your kids the same way. Make sure they have the right clothing, shoes, communication devices, and information about what to do in an emergency. Allow them the freedom to run around without fearing the dangers they have to worry about in the big city, but make sure they have the tools and knowledge to keep safe.

Chapter 5
Equipment

Most people who have dreamed of owning a weekend place have imagined themselves riding around their acreage on a John Deere tractor or a utility vehicle. Farm toys bring out the child in us all. Whatever brand you select, the key to controlling upkeep time and the costs of maintaining your land is to purchase the right equipment. Your agent will be able to help you understand your needs based on the size of the property you are buying. Helping clients shop for the perfect equipment for their new property is possibly my favorite part of the buying process. Here are some options for acquiring the right equipment.

Option 1: If you like the way the seller maintained the property, it is very common to buy the ranch's equipment with the purchase. The sellers no longer need the equipment they have been using to maintain the property at the level you fell in love with when you decided to buy it. If the numbers make sense, grabbing the seller's equipment at the point of sale is the easiest and most seamless way to stock your place with the right equipment.

If you decide to go that route, make sure your agent asks the seller's agent for a list of year, make, model, and engine hours. And be sure to take pictures. You can then take them to a local tractor shop or equipment store and get a pretty good idea of their value. **If you e-mail the list to me, I would be happy to give you a general idea of the value.** Once again, the convenience factor of already having the equipment on-site is worth something. I find that sellers are reasonable with their prices if you are paying them what they generally want for their property. Sometimes, they don't have the right small equipment, but their big stuff is just right. In Appendix B, you'll find a starter list of small equipment I recommend. It's also printable from our website. The key is that the equipment is simple and safe for users and that you know what to expect in terms of longevity.

Option 2: Another way to handle the equipment your property requires is to not buy or keep much equipment on the property. Instead, find contractors who will do all the large equipment jobs on a calendar schedule and use their equipment. That option isn't as great financially since you end up paying much more for the labor. You also wouldn't have any of the equipment to tinker on when you are at your ranch. The only time I would recommend this option is if you have just bought a place that comes with no equipment and you are not sure about what you need or want to keep maintained on your ranch.

An alternative to option 2 is to allow contractors to do the larger projects or all the projects, but you buy the equipment and allow them to charge you an hourly rate for the labor.

Commercial-Grade Equipment

Nearly every time I come onto a ranch after someone has owned it for a while, I see a pile of low-cost tools sitting in a corner collecting dust. Don't go out the weekend after you buy a ranch and get a bunch of cheap equipment. If I could write this advice on a piece of paper and attach it to everyone's forehead, it would say this: Only buy commercial equipment! No lawn mower at Lowe's or Home Depot is ready for ranch life. Yes, they work fine for small residential properties, but for continual use on a ranch, they do not hold up. Make sure each piece of equipment you buy is commercial-grade and heavy-duty. Ask a salesperson if you're not sure. Commercial equipment costs twice as much as residential models, but they are manufactured with metal components, not cheap plastic. I'm a huge fan of John Deere or Kubota but have used many other brands and like them as well, just as long as they are commercial-grade.

We offer a ranch set-up consultation that includes an assessment of the equipment we recommend for your specific property. I'm happy to meet you at your property, look it over, and consult with you on what

to buy. We tend to look for used equipment in good condition whenever possible and buy new for high-use, high-frequency, lower-cost equipment.

I buy an array of Stihl hand tool products for each of my ranches. In Appendix B is a list of the model numbers I prefer. They include items such as weed eaters, chainsaws, pressure washers, pole saws, hedge trimmers, and more. We have relationships with our local equipment providers and are happy to help any of our clients with their first line of equipment purchases.

I would love to tell you exactly the right size tractor and mower for your place, but it does vary based on several things such as the property size and your desired use, which can change the price immensely. The smaller equipment recommendations are almost always useful, but when you get into the big, heavy-duty tractors, mowers, and utility vehicles, it becomes hard to give advice without seeing the property and hearing your desired uses.

We went over this in the chapter on safety, but I want to mention it here again. Nearly all tragedies on ranches are a direct result of improper equipment use. People get hurt from not being cautious, from not paying attention, or from using a piece of equipment they know nothing about. I'm not suggesting you be afraid of or avoid buying equipment because it can be dangerous, but it is very important to understand its limitations, maintain a safe speed, and pay close attention to what's going on around you.

Chapter 6
Must-Haves for Your Ranch

Often when I meet with clients, I jokingly tell them I know what they want more than they do because we have seen, thought through, and fleshed out endless ideas for how to be efficient and effective on ranches. We know what stuff is useful to own and what stuff is not worth the hassle. This chapter is all about what my high-end clients love to have on their properties.

A Utility Vehicle

If you buy a place with some acreage, you and your guests will benefit from owning a utility vehicle simply for the convenience of getting around. I own one and consider it the best purchase I made this year. I've brought mine to other people's properties, to show potential buyers a new property, and to check the mail. I even take it to West Texas to use while hunting. Be prepared for sticker shock but also be informed about selecting the right one.

I have owned a couple different brands for my properties and know which one will best fit your needs based on your desired use. If you just want a sporty luxury vehicle—the Maserati of off-road vehicles—you should consider a Polaris Ranger 900 or 1000. They are sporty and comfortable but, like a Maserati, require more maintenance and aren't as dependable for hard work. Our company owns six Polaris Rangers, and we love them, but the yearly upkeep cost is significant. I'm still a huge fan of them, and I think it's worth it if you've got the budget. Currently, Polaris utility vehicles come with a six-month warranty, and if you don't abuse them, they will last.

If your hope is to enjoy nature without the roar of a motor, another option is the Polaris EV (electric vehicle). It's great for roaming around your land and sneaking up on game but not great for work or long, hilly rides. In my experience, the battery will last around 10 miles, not the 40 miles that Polaris advertises.

I also bought a comfortable and dependable Kawasaki Mule Pro-FXT. It's a crew cab, so it sits six people comfortably and has a cargo bed for bags of corn, a Yeti cooler, and more. It comes with a great warranty, which tells you the company expects it to last. I have worked it hard in West Texas, driving more than 100 miles in a day on extremely tough terrain with five large adults, corn, and various gear on board. So far (knock on wood), it has had no breakdowns. The sticker price ranges from $13,000–$21,000, but they are hard-working, durable machines.

There are other brands a little bit cheaper, and there is always the option of a used piece of equipment. But my warning about going that route is that you don't know the depth of water it's been in or how it was taken care of. I highly recommend buying a new machine as soon as you get your place and then maintaining it so it will last a long time. As a disclaimer, these vehicles must be treated with respect, and adult supervision is highly recommended. Accidents happen all the time on off-road vehicles, and they can result in injury or even death. So be careful, be wise, and be safe.

Giving suggestions about equipment is one of the favorite parts of what I do. So please don't feel bad about calling me and asking for advice or hiring us to help you get exactly the right equipment for your property. For my real estate clients, this is a perk I will gladly throw in as part of my service as your realtor.

Skeet Range or Shooting Range

Ninety percent of the properties I've worked on now have some sort of shooting range. The most fun is skeet or sporting clay. If you plan to host at your ranch, skeet shooting is an excellent way to blow off steam after a hard week of work. Even if your skeet range doesn't get used often, it is not very expensive and is a ton of fun. I have great recommendations on installers, layout, quality mid-range-priced shotguns, and everything in between. We can also show you pictures of some options for laying out your skeet range.

Another fun shooting feature is a pistol or rifle range, which usually consists of a large dirt mound covered in grass with various metal interactive targets. Adding a pistol or rifle range is a very fun addition to a property and a great place to teach your family and friends gun safety and responsibility.

A Firepit

So many of the daydreams people have about the optimal way to spend time with family and the amigos include a firepit. Of all the outdoor fun activities my clients set up, their firepits probably get used the most. Your firepit can be as simple as stacking some stones and getting Adirondack chairs from Home Depot or as elaborate as granite tables and copper pits with stone inlays. Either way, it's a great use of your time and space. It's also a great spot for you and your family to connect, slow down, watch wood burn, and get closer.

Feeders

One of the many gifts of living in the country is increased access to wildlife. If watching wildlife is on your priority list, the best way to get consistent, continued results is to have a game feeder you can see from your house. It's so fun to see all the different critters eat rather close to you. Add to your must-have list a decent set of binoculars, and sit back and enjoy God's creation from your back porch. If your plan is to get a wildlife exemption, a gravity-fed feeder can be part of your application process.

You get what you pay for with feeders, so a mid-range galvanized feeder is a good option. I like All Seasons feeders, but there are many with a similar design that are good. I manage hunting on a 5,000-acre ranch and have found the cheap drum feeders are good for about two years, but the All Seasons brand will last a good 10 years. The main difference is the fasteners used to hold them together. Spend a little extra—$600 or so for a good one—and you won't have to mess with it for a long time.

A Housecleaner

It may seem funny to include a housecleaner on a list like this, but as I mentioned in a previous chapter, I guarantee you the annoyance of cleaning when you arrive and deep cleaning before you leave makes owning a weekend place a pain in the butt. Go to your local country store or look online on Facebook or the Chamber of Commerce website and find a cleaning person you can text a week before you're going to arrive or give a Google calendar full of scheduled cleaning times. Let them know you expect the place clean when you get there, and then ask them to do a deep clean after you leave. It's worth every penny. As previously mentioned, a common reason people decide to sell their property is because of the headache of all the housework when they arrive and again before they leave. If you can afford it, get somebody else to do it. Remember the cost of your time, and remember the reason you bought the place.

A Barn Cat

Once you've bought your dream tractor and utility vehicle and set up your skeet range, the next project is to keep all that equipment in good working order. You might be surprised to learn the biggest preventer of that is mice. It's the country, and yes, there are mice. Adopt yourself a couple of male cats so you don't end up with a bunch of kittens, and feed them enough so they will be snake and mice hunters when you leave them during the week. If you don't love snakes, cats are great. I'm cringing a little as I type this because I'm not a cat lover, but we have an outside cat that's rarely on our porch and never in our house and protects our equipment. It's a no-brainer to get a couple of cats and leave them to run roughshod on the rodents and reptiles in and around the barn. They'll be some happy felines.

Outdoor Games

Don't be the stereotypical weekender who buys an amazing ranch in the country and then goes there to watch CNN. Get the family outside, catch a fish, shoot a gun, and play some games. This is how memories are made. Whether it's a frisbee golf course, a good corn hole set, or a driving range to help you improve your golf swing, add it and use it. My family loves frisbee golf, adult slip-and-slides, and ping-pong. These games get a lot of use around our place.

We have built, found, or made all the items listed here, and we're happy to help new ranch owners get their outdoor activities set up. Make your place a home away from home. Make it a place where you recharge and improve your physical and mental health. Make it all you dreamed of. I'm here and happy to help you.

Chapter 7
Investment

Here is a common question people often ask when I'm showing them a property: How do I make money on my property? Passive ranch income is a great idea but unlikely. Covering your costs with leases is possible, maintaining your ag exemption with a cattle tenant is probable, but moneymaking in real estate is about one thing: buying right—buying right—buying right.

I personally own all types of investment real estate—residential, commercial, farm, and ranch. At the time of this writing, I have several commercial buildings, eight or so houses in Brenham, Texas, and several farm and ranch properties in Chappell Hill. I live, eat, and breathe real estate investment. And let me tell you why. I have what many would consider two to three full-time jobs. My amazing team and I manage large weekend ranches, I sell real estate, and I invest in real estate. The easiest money I make from these three jobs is from investment.

Real estate investment in the right area at the right time is low-risk and high-reward. For instance, I'm located one hour from Houston, one and a half hours from Austin, and one hour from College Station. Historically, real estate is a safe bet here. The appreciation of land has far exceeded inflation rates, and the ROI has been excellent. I also own some stocks, but the major difference between that kind of investment and real estate is that I can't go hike on my stock; I can't go hunt on my stock; I can't take friends to my stock; and historically, my earnings on land have far exceeded the cash-on-cash return in the stock market.

I mentor many people in real estate investing. It's my goal that all my friends and anyone I meet who's interested in real estate succeed in the real estate investment business. I love watching people succeed. At the risk of stating the obvious, I have to make one major disclaimer: I can't guarantee your results, and just like in any market, there's an ebb and flow in real estate investment. The other factor to consider is what your ultimate goal is for the property. If you're looking for your family's dream place *and* looking to make it the best investment of your life, that is a conflict of interest. Even though I expect you would do well if you find the right place, I make no guarantees.

If you have the money to buy your family's dream place, don't be overly concerned about the next section. Sometimes you've got to take off your work boots and put on your flip-flops—or socks and Crocs, in my case.

Exit Strategies

There are a few parameters I look at when buying real estate. The first is multiple maturity options, or exit strategies. That can show itself in rental potential, appreciation, depreciation for tax benefit, leasing land, and so on. When I look at a piece of property, I always think about its possible uses. What are the sellers not highlighting that this place offers? Can it be split? Can it be developed? I'm naturally an overly optimistic guy, but I always try to list the potential uses of a property for the buyer. It's something I do with my clients when we're looking at land. The conversation gets more animated when I ask something like this: What kinds of additional uses does this property have that would interest you? Airbnb? Corporate event ranch? Few things get an entrepreneur more excited than getting to own a ranch *and* offsetting many of the costs. But for about 80 percent of people, the reality of a limited availability for the owner, not to mention the extra brain space it takes to execute and manage it, discourages them from exploring these other revenue streams. Still, it's an option. As I write this, I'm helping some clients negotiate a six-week lease on their horse barn for a polo event with income that will offset their costs for eight to ten months this year. Now that's exciting! They listed their amazing barn at the Houston Polo Club, and boom! Here they are.

The ranch rental or wedding venue is a trending option and one that generates a lot of interest. The idea is popular because renting out a stellar property for weddings or other special events can offset upkeep expenses. Who doesn't want to rent a spectacular old barn restored as a dance hall? Who doesn't want to throw a 50-year anniversary party for their parents by a beautiful lake? I've met several people who build their dream property and then turn right around and create an Airbnb listing. They spend a few thousand dollars to hire a marketing agency to get direct-to-consumer marketing to high-end businesses for weekend retreats. The big hole in the Airbnb industry is properties for large group rentals, and ranches can fill that need. The negative, obviously, is that your ranch may get booked up and you'll only get to come every third or fourth weekend. But the ability to run your ranch as a business and claim losses as you build has many benefits. I've seen people have a lot of success with this model. I'm happy to meet with anyone who would like to brainstorm rental ideas.

I've talked a little bit about cattle leases and hunting leases, but I'll mention it again here since it's a good way to offset taxes and create a very small income. The going rate for great cattle land is about $30 per acre per year. A great hunting place is around the same per acre. Unless you go extreme on the income production such as building an event center or something similar, count on the appreciation of your land to be your moneymaker.

An often overlooked reward that justifies owning a ranch property is allowing people, organizations, and friends you trust to use it. My family and I are active in our local church, and I am involved in many ministries. I love being able to offer my property to host groups that would otherwise never get to enjoy a place like mine. This is not income production in the traditional sense, but making your property available and serving others pay much larger dividends than a cattle or hunting lease. It's a significant deposit in the bank of your well-being and happiness. Plan on being the person you always wanted to be and allow your ranch to be a way for you to give your time, talent, and resources.

I also think of my friends, the Camerons. They own a great property in Chappell Hill, Texas, and I have watched them make it available to people and groups they believe in, even if they don't know them. It's not making them any money, but the joy of getting to know new people and the pleasure of watching their place being used for Kingdom work are more fulfilling than any income. There are times when selling their property and moving closer to Houston would have been easier, but it's exciting for them to watch people playing tennis on their court, swimming in their pool, and dancing in their barn. Income on your property is great, but the impact you make through your property is greater.

Investment Story

As I write this, I am working with a family who has a great plan. They are looking for a place that meets many of their wants but has the infrastructure already complete and is in a place where they expect appreciation. The husband's main goal is to own a place for 10 years for free or close to it. He wants to sit on a place and enjoy it with his family as the market climbs and sell it when it's time to pay for his kids to go to college. These opportunities are possible. The main factor and main message I reinforce with him is to leave it alone as much as possible. Adding too much is the way to lose.

Chapter 8
Wildlife Tax Exemption

Most people buy land in a rural environment so they can enjoy the quiet surroundings and the wildlife that abounds. While you are enjoying all of this, why not get a tax exemption at the same time?

In 1995, Texas voters added an amendment to the Texas Constitution that open-spaced lands devoted to wildlife management should be assessed based on their productive value. In simple terms, that means your property may qualify to be taxed based on its productive value rather than its market value. That could save you a significant amount of money in property taxes. Most ranches have an agricultural exemption in place. You can go to the county's appraisal district website and search an address to find what is in place and what the taxes are. Your real estate agent will do that for you, but make sure to look at the site before closing on your place.

Most country properties are in traditional agricultural exemption areas. That means they host cattle, hay, or other animals. But a lot of properties don't fit that model, and you may not want to mess with it. You can switch from traditional ag to wildlife and make adjustments to your place that invite and hold more wildlife. Most of my clients choose this route. While both exemptions provide the same tax benefits, a blend of the two is even better. The main advantage of being categorized as traditional is that fair-use tools are tax-free, whereas they are not for wildlife properties. That means if you keep a small hayfield or buy a couple of cows and keep a small cattle field, your farm purchases are tax-free with a Texas ag or timber number that you can get online in five minutes. I have an excellent in-house team member who does all our land tax work. He can meet you on your property and get all the paperwork completed and maintained each year. It may be the best money you spend.

When considering applying for a wildlife management appraisal, it's best to try to acquire property where the process for agricultural appraisal has already been started. According to the Tax Valuation for Wildlife Management website, "Only properties that are currently being appraised as agricultural lands or timber lands may convert to appraisal based on wildlife management."[1] Since the appraisal is based on a seven-year cycle, you would do well to buy land that is already in the qualification process. The cycle applies to the land, not the owner of the land, so you are well ahead of the game if you purchase a property already in the rotation cycle.

Requirements

The requirements for wildlife management are not difficult. You must furnish a wildlife management plan to your appraisal district and provide proof of ownership, a description of the land, and the current use of the land. You will be required to indicate your goals for the land, including which indigenous wildlife you plan to manage and how you intend to carry out that plan.

There is plenty of information on how to go about making a plan. If you are a do-it-yourself person, the Internet is a veritable gold mine of information. The Texas Administrative Code provides various links that explain the intent and procedure for wildlife management.

The FAQ section of the Tax Valuation for Wildlife Management website (https://tpwd.texas.gov/faq/huntwild/tax_valuation.phtml) is an excellent source of information for any questions you may have. Another site many people find useful is the Legal Summary of Wildlife Management Use Appraisal (https://tpwd.texas.gov/landwater/land/private/agricultural_land/legal-summary.phtml).

On the other hand, if you have purchased property in the country in order to just relax, you may want to pay a little more and have your real estate agent help you acquire acreage already in the exemption application process. Then you will probably want to hire someone who understands the

1. Tax Valuation for Wildlife Management FAQ, *Texas Parks & Wildlife*, https://tpwd.texas.gov/faq/huntwild/tax_valuation.phtml

ins and outs of setting up a wildlife management plan. Again, your neighbors will know what wildlife management plans work best in your area. There are many options.

Agricultural workers are another helpful resource for what kind of wildlife is best suited to a particular area. If you have hired a ranch manager, he or she can research and help set up habitats and shelters for any wildlife you choose. We at Brazos Ranch Group are always available to help. We care about the maintenance of the land and the wildlife, and we are committed to helping owners make their land meet all their expectations while maximizing its tax shelter and appreciation potential.

About Taxes

Jen and John have owned their ranch for 10 years. When I met them, they were interviewing me to sell their ranch. I listened to their pain points—the things that made them want to sell. They wouldn't stop saying how much they loved their ranch but how much the work consumed them. I persuaded them to wait six months to sell it and let me run it for those months. The most interesting thing was that they had traditional ag tax valuation, but their most annoying problems were maintaining the hay lease: finding people who said they would take it, getting them to show up, and dealing with the cattle lease tenant. He was overgrazing their place, and his cows had eaten their gardens and broken their fence. That was the last straw. One cow had decided to relax in their pool when they arrived one weekend. We were able to quickly transition them to a wildlife tax valuation, add more things that attracted wildlife, and get all hay and cattle people off their ranch. That was two years ago, and we no longer manage their ranch. They hired a guy to run their place, and they still own the ranch. They send me real estate clients often but always tell them to be careful or I will persuade them not to sell.

Chapter 9
Regrets

I have seen clients implement some amazing ideas worth modeling, but I've also seen a few plans over the years that sooner or later become regrets. You want owning your property to be a great experience, and streamlining and systematizing the way you manage your property, including the equipment, goes a long way to ensure that owning a weekend ranch is painless and brings you and your family years of enjoyment.

Regret #1: Homeowner Models vs. Commercial Models

The number-one regret is starting with the cheapest equipment. I mentioned that earlier in the chapter on equipment, but I honestly can't say it enough. Almost every time I visit a property for the first time, I check out the barn, and inevitably in the corner under a bunch of dust are the seller's original equipment purchases when they first bought the place. They drove to the local home building store and bought the homeowner models instead of the commercial models.

On our website, I provide a list of items I buy for every ranch. It is well worth your time and money to start with the right equipment. You might be able to purchase a chainsaw for $99, but you will not be using it in six months because it will be broken. If you buy a zero-turn lawnmower at Lowe's for $2,000, it won't last. Even if you buy the high-end model of the homeowner models, you will regret it. Ranches demand durable equipment that can handle heavy use, varying terrain, and vast acreage.

Regret #2: Superfluous Things (aka Ranch Hoarders)

Regret number two is the temptation to fill your weekend home with unused things from your primary residence. It's hard to throw things away. You never know if your great-aunt Sally's tea set will come back into style in 40 years. But let me give you some valuable advice. If you didn't use it in your primary residence, don't bring it to your weekend ranch. All that broken or unwanted stuff stacked in your house or garage should go to the Salvation Army. Your ranch won't be a place where you can relax if it looks like an overfilled storage unit. It seems obvious, but let me risk being redundant. Make your ranch house comfortable and practical. Ain't nobody want to be on the next episode of *Buried Alive Ranch Hoarders*. So donate it!

Regret #3: Trying to Do It All Yourself

Some people have trouble with regret number three. Yes, you own and run a successful business in the city, and yes, you are capable of successfully running your weekend place. But why? What's the value of an hour of your time? Would you book a vacation at a beach resort if you had to rake the beach, clean your casita, and chase iguanas out of your room—even if you do have an amazing talent for it? No! So don't mess up weekends at your property by overdoing it. Save your favorite tasks for you, and pay $20 per hour for a great person to prep the stuff you don't like doing. Budget accordingly and, if you have to, wait on the skeet range. But get help. Get. Help.

Chapter 10
Plugging In

One of the greatest resources for retaining a passion for your weekend property is building a sense of community in the town where you buy. I encourage you to find ways to participate in events that interest you. Attend community classes on topics such as gardening, fruit orchards, or healthy trees and meet people who have similar interests. Go to small-town stores and meet the owners. Find a church that jibes with your beliefs.

Our small town has museums that need volunteers, an historical association, numerous church opportunities, and a wildlife association. The point is to go to town gatherings and meet the locals. Tell them where your place is, ask them questions about the area, and get information on the history of the town. Locals love sharing their town's history.

Meet your neighbors. Get on the appraisal district website listed for the neighbors who live near you. Click on the maps link and view the public information listed for the neighbors who live near you. Reach out and invite them to your new place. I remain close friends with most of my clients, and I love watching their places mature and seeing what they decide to do with them. I love inviting them to my home or meeting them at a local bar or restaurant. The small town country way of life—meeting neighbors, jawboning about the weather—is contagious.

I get it. The main goal is having a place to unplug from society, but there is a big value in knowing people in your new community. Allowing people to come on your place to catch fish with their kids or have a meeting will give you so much more fulfillment and joy in owning your property. That doesn't mean people have to come out every weekend or that you need to attend every event you're interested in. Beware that you don't overschedule yourself. But it's great to have one or two occasions a month that require your presence. The rest of the time, you and your family can veg out and enjoy the peace and quiet.

BRAZOS RANCH
GROUP

Chapter 11
Let's Do This

Owning a weekend ranch is a great decision. Now that you're a little more informed about the process, it's time for the real lessons to begin.

We would love to help you find and buy the perfect place. We would be honored to help you get your land, set up the management of the place, and plug you into all the people and information we have. I wholeheartedly know my team is the group to serve you. Because we both sell and manage ranches, we are the best choice to help you understand upkeep processes and estimates for each property we show you. I would like to meet you and begin the process, and I'd love to be your go-to guy after the sale. My clients still regularly text me with questions, and I'm always happy to help them any way I can.

Do you own a place and want to sell it? Don't let that stress you out. Our team will come in and handle needed repairs so you can get your property listed and sold. We offer what no one else can, so you will experience the next level of excellence in cleanup and sales prep. We can also maintain your property while it's on the market. I want every showing to present your place at its best. I maintain several of my clients' listings and have had great success selling many properties quickly and for a great price. First impressions are key.

Don't wait. Make your family's dream a reality. Let's get started on what I believe will be one of your best life decisions. Start living your dream today.

Sincerely,

Matt Montgomery

Bevers Real Estate, Inc.
281-795-4471 (Matt's cell)

Brazosranchtx.com
979-830-1180 (Bevers Office)

Appendix A: Inclusive Contractor Needs Form

You may not need all the information in this appendix, but many property owners do. When a person buys a property through us or hires us, they have access to our qualified contractors, and we will fill out this page for them.

Alarm company
Appliance repairman
Cable company
Cattle guy
Electrician
Equipment sales company
Equipment service company
Ferrier
Gate company
Handyman
HVAC company
Internet technician
Local emergency
Local police department
LP company
Neighbors' contact information
Phone company
Plumber
Pond management
Pool company
Septic inspection or repair
Tree company
Veterinarian
Water well company
Yard company

Appendix B: Small Equipment List

In this section, you'll find a list of must-have small equipment useable on almost every type of country property. These are high-quality products, each with their own unique use. I highly recommend them to make your life much easier.

Model	Line	Description	Quantity	List	Net	Amount
HT133	STQW	Pole Saw	1	$649.99	$625.99	$625.99
MS250	STQW	45.4cc 18 Chainsaw	1	$359.99	$299.99	$299.99
FS91R	STQW	28.4 Cc Trimmer	1	$329.99	$299.99	$299.99
HS46C	STQW	22in E2s Hedgeclipper	1	$369.99	$369.99	$369.99
BG56C	STQW	Ez Start Hand Blower	1	$179.99	$179.99	$179.99
RBA400	STQW	Stihl® Pressure Washer 2700 PSI	1	$599.99	$599.99	$599.99
					Invoice Total	$2,375.94
					Sales Tax	$196.01
					Grand Total	$2,571.95

About the Author

Matt Montgomery is an outdoorsman and hunter whose passion is land sales and stewardship. After graduating from the University of Houston, he returned home to turn his love for the outdoors into a thriving career managing ranches and serving as a real estate agent specializing in farm and country properties. Throughout his more than 10 years of ranch management experience, Matt has bought, cleaned up, and sold real estate, cultivating a deep understanding of pricing and market trends and managing more than $100 million worth of weekend country property in the Brazos Valley. Matt serves with his wife and three kids at Brenham Bible Church and sells property with Bevers Real Estate.

CPSIA information can be obtained
at www.ICGtesting.com
Printed in the USA
BVHW090907190319
542974BV00006B/14/P

9 781632 963246